Name: _____

MAGIC NUMBERS
STUDENT BOOK 2

A. A. Sarkiss

Printed by CreateSpace, An Amazon.com Company

Available from Amazon.com, CreateSpace.com and other retail outlets

Revision

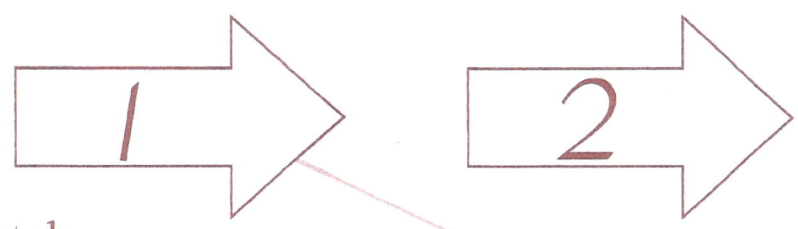

Count and match.

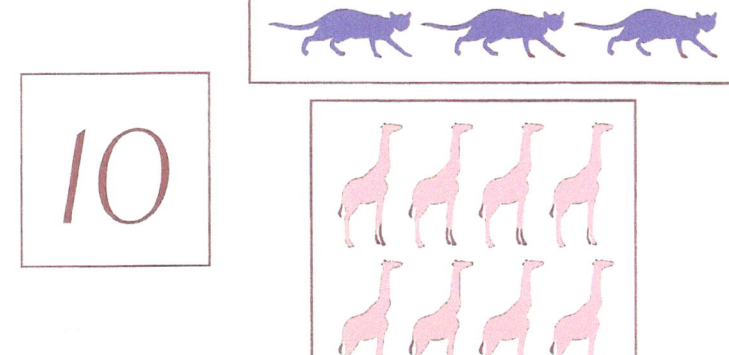

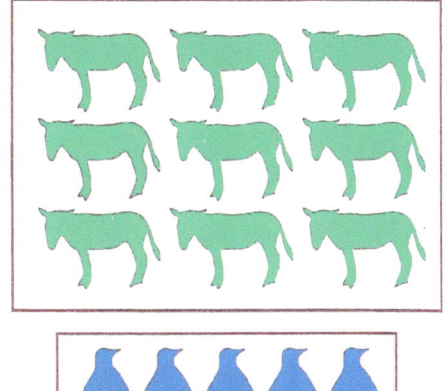

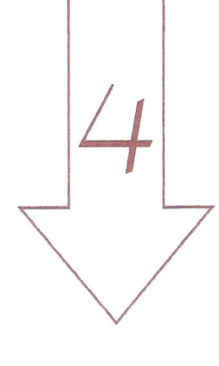

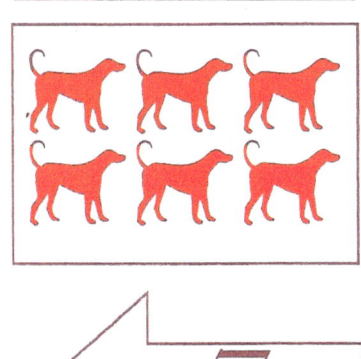

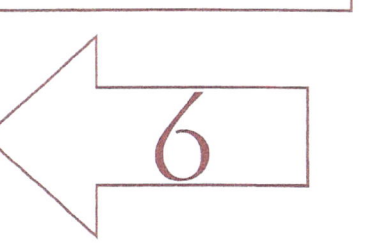

Revision

Write the number before or after.

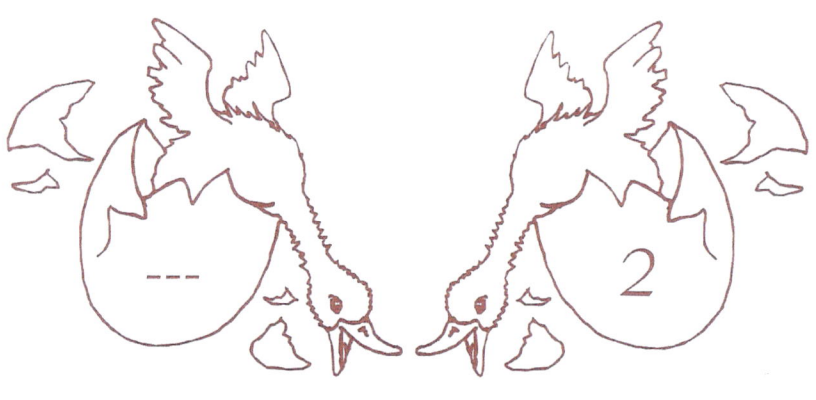

Revision

★ Circle the greater number.

- (3)   2
- 9   7

- 6   8
- 10   8

- 4   5
- 1   2

- 2   6
- 7   5

★ Circle the smaller number.

- 10   (9)
- 4   2

- 1   3
- 6   7

- 8   7
- 3   5

Revision

Add, write and match.

★ | $4 + 4 = \underline{8}$

★ | $3 + 2 = \underline{}$

★ | $6 + 3 = \underline{}$

★ | $5 + 2 = \underline{}$

★ | $3 + 1 = \underline{}$

★ | $4 + 2 = \underline{}$

★ | $1 + 2 = \underline{}$

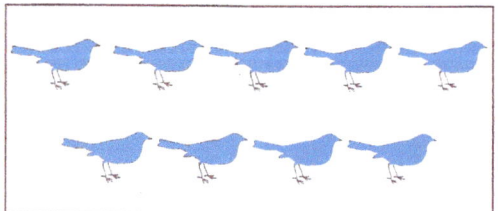

Shapes

1 - Match.

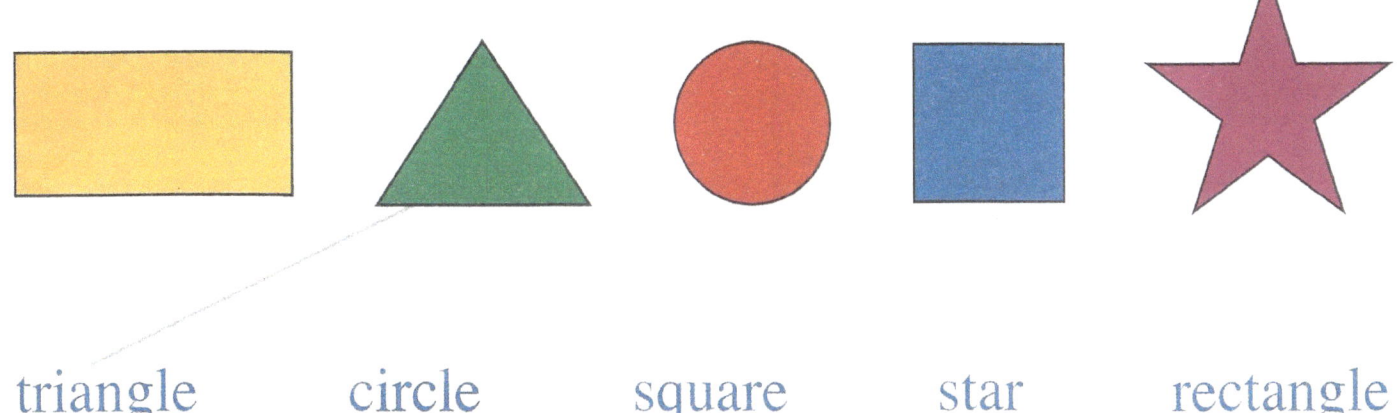

triangle   circle   square   star   rectangle

2 - Circle the same shapes.

Count and match.

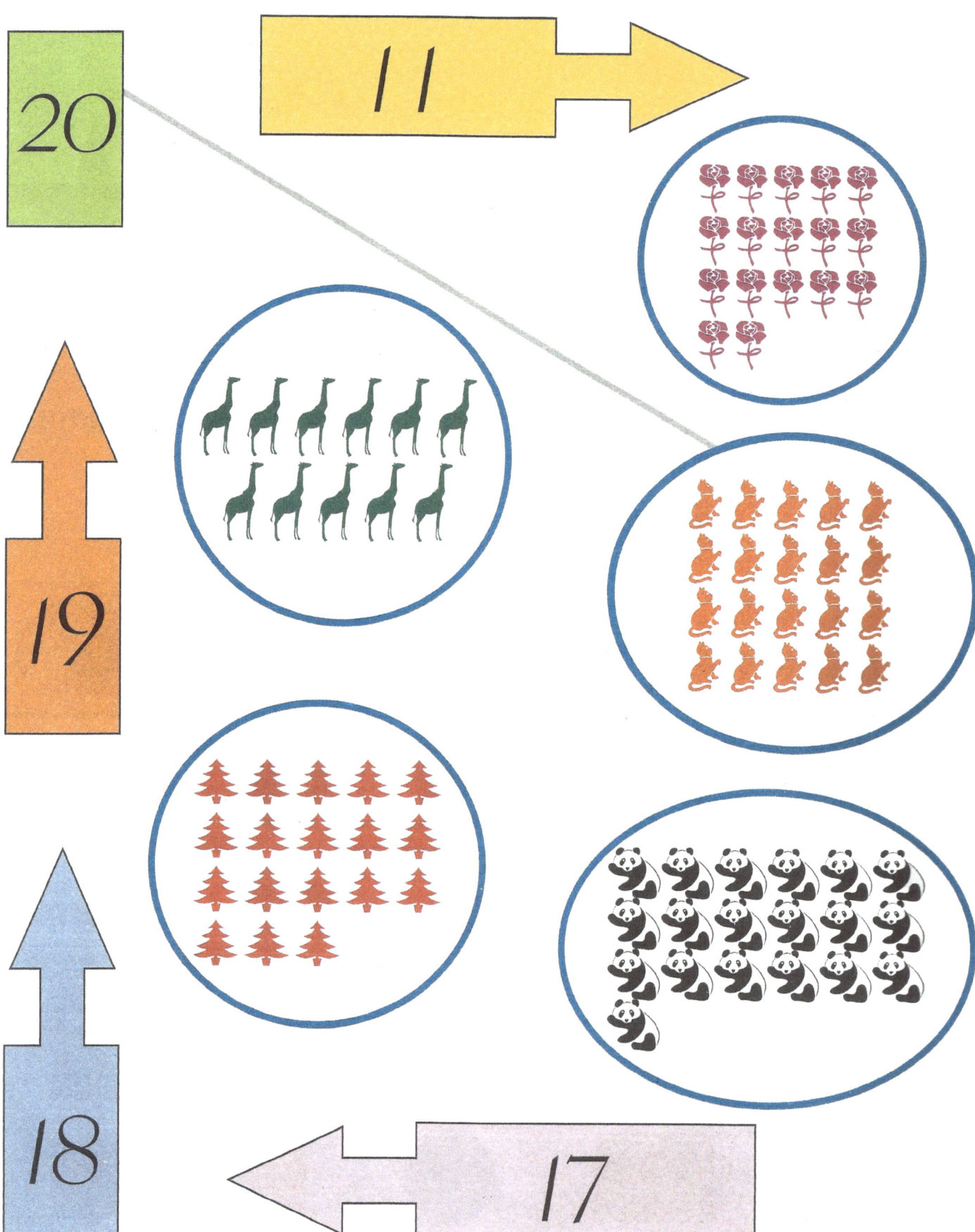

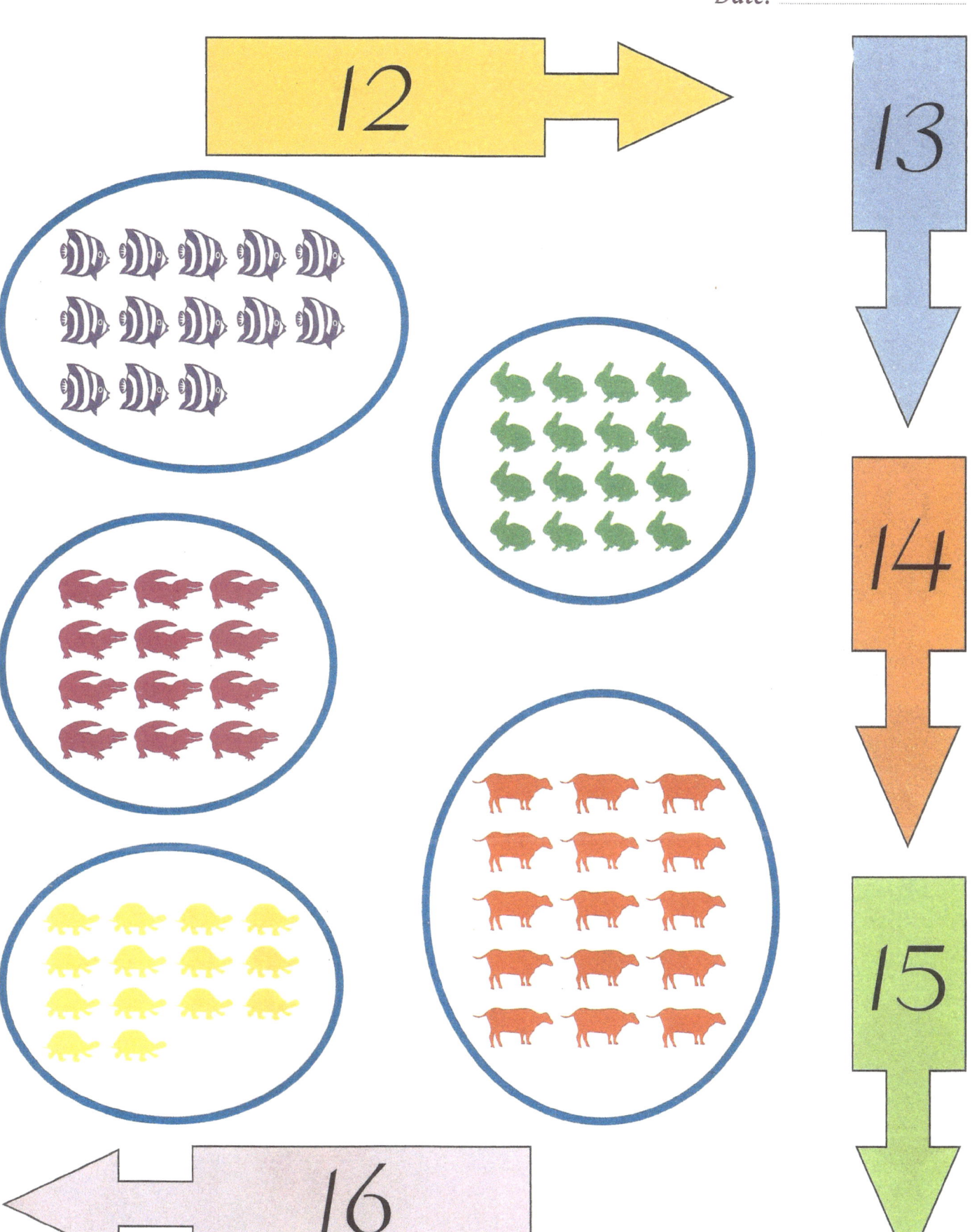

7

Fill in the missing numbers.

    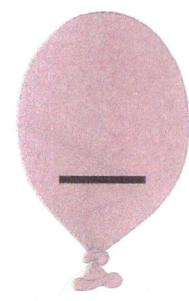

  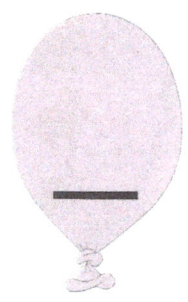  

   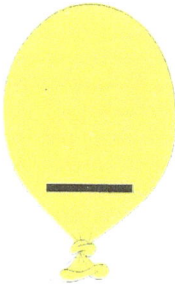 

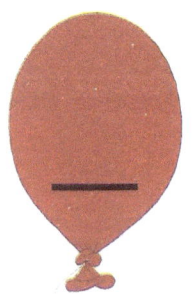    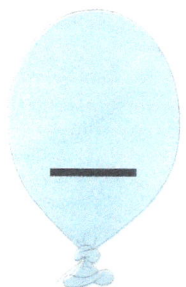

Write the equation.

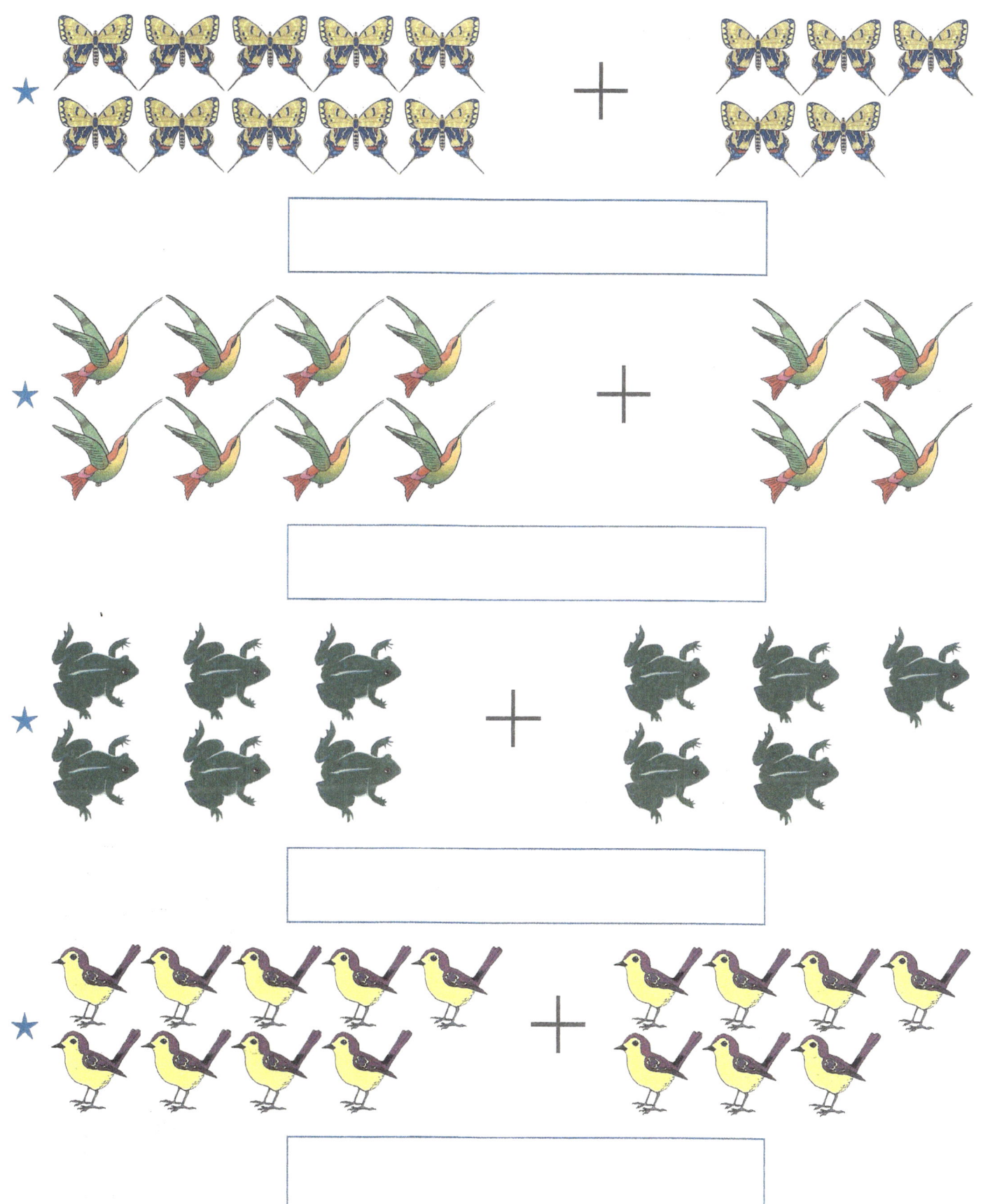

⭐

⭐

⭐

⭐

Match objects of similar shape.

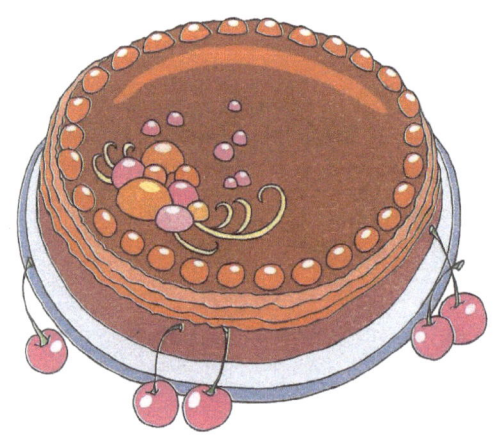

Colour according to shape.

pink	green	red	yellow	orange	blue
⬭	▲	★	●	■	▬

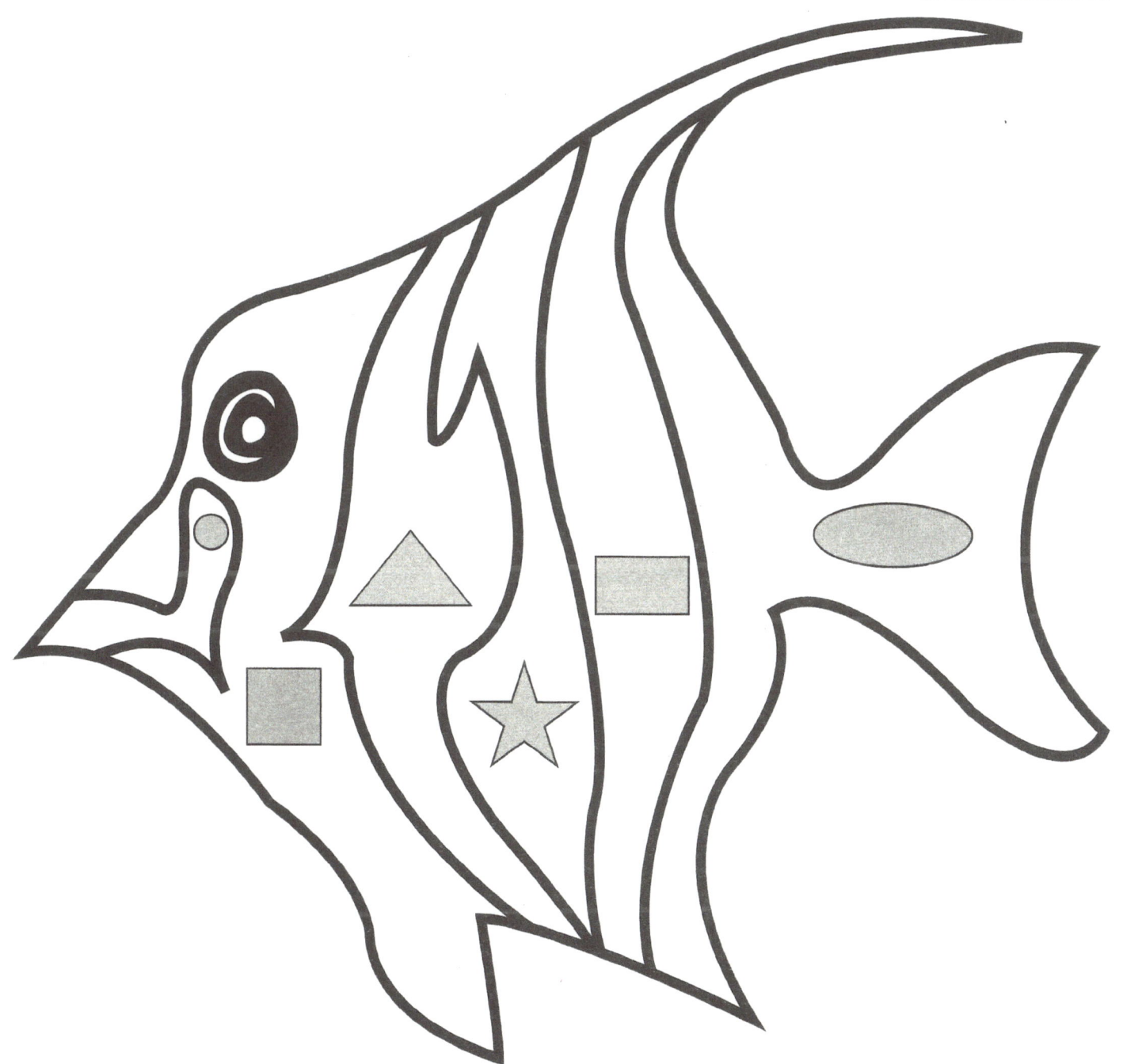

Count and match.

21

30

29

28

27

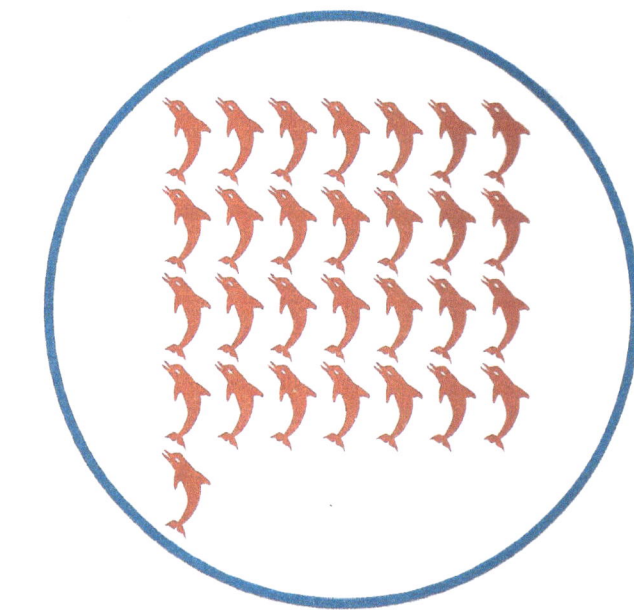

22

23

24

25

26

Subtract and write.

3 – 1 =

5 – 2 =

   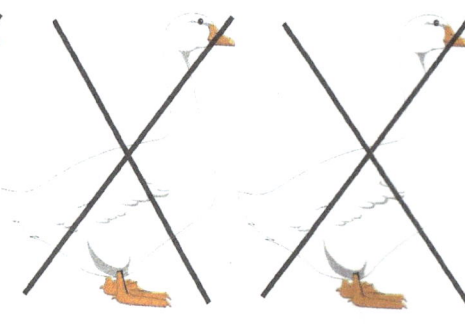

7 – 3 =

8 – 2 =

Match each object with a similar shape.

Add more to the group to make the number.

Write the equation.

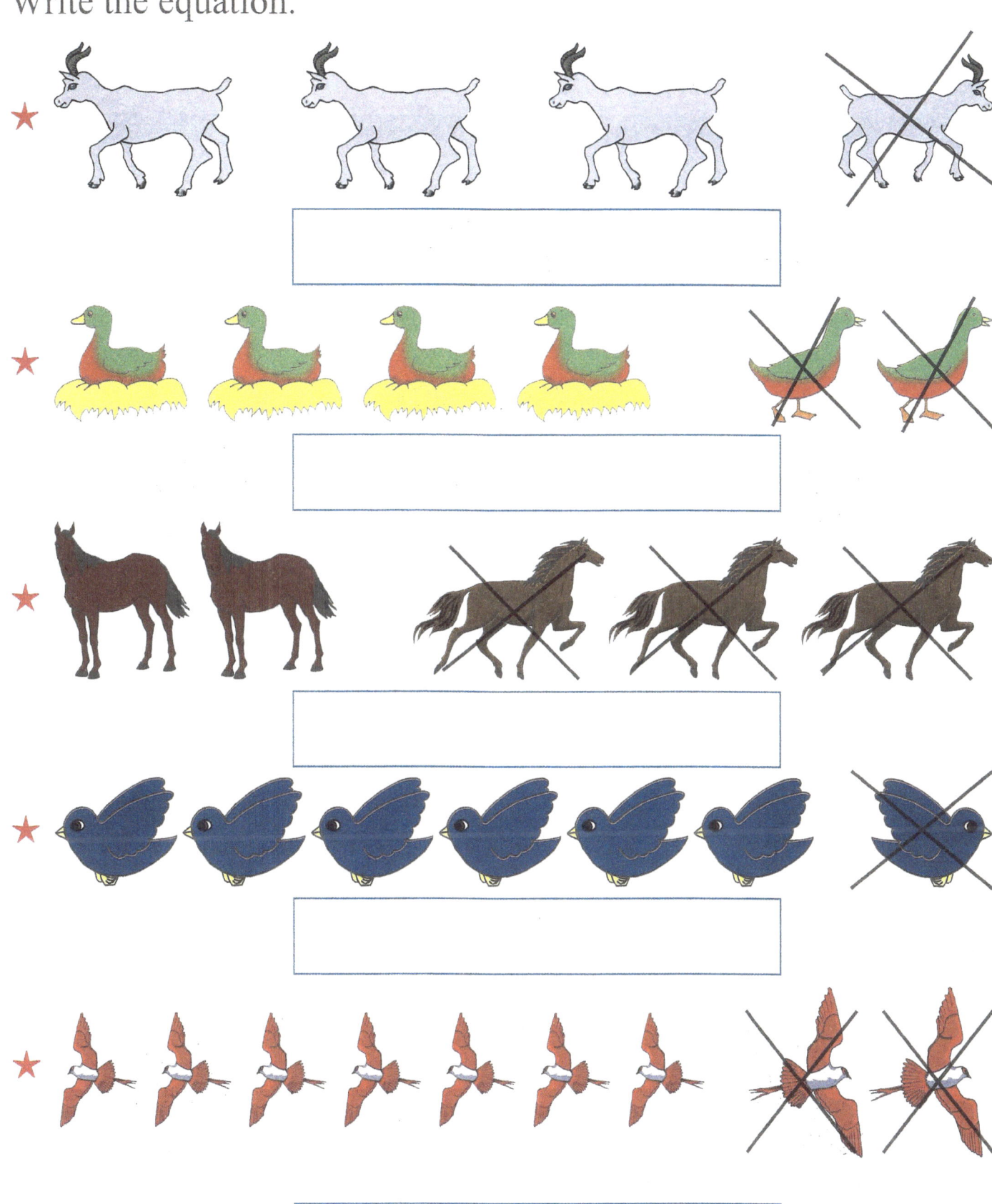

Find four differences.

A

B

Add more to the group to make the number.

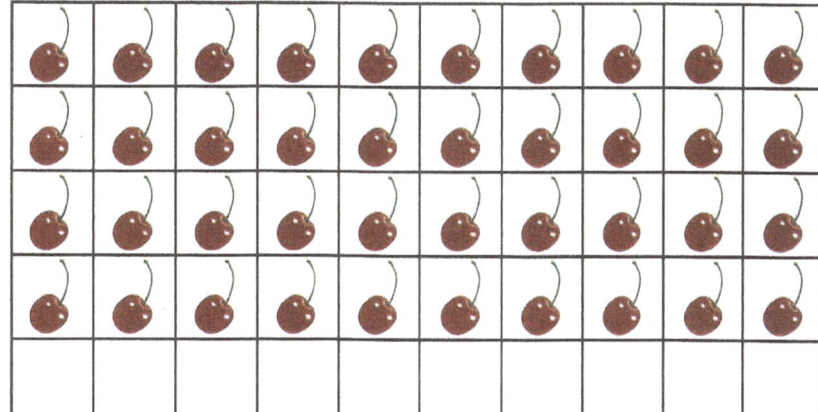

Money

Copy the pattern.

1.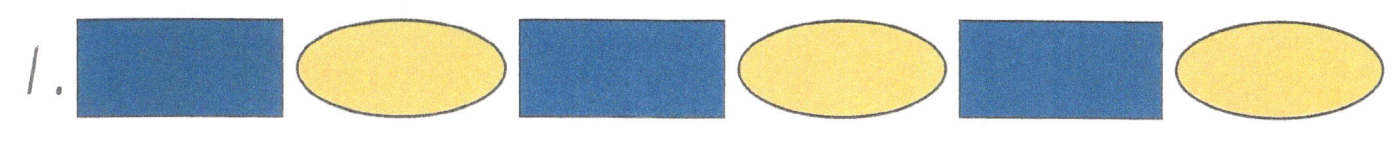

2.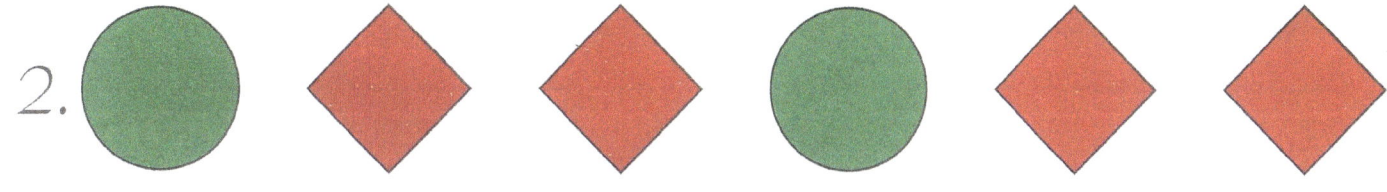

3.

4.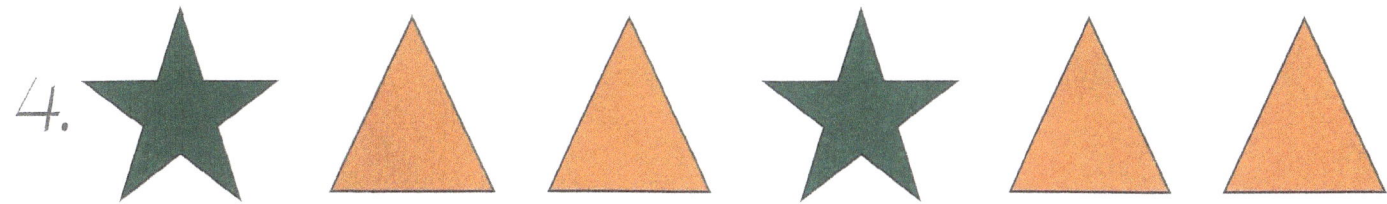

Add and circle the correct number.

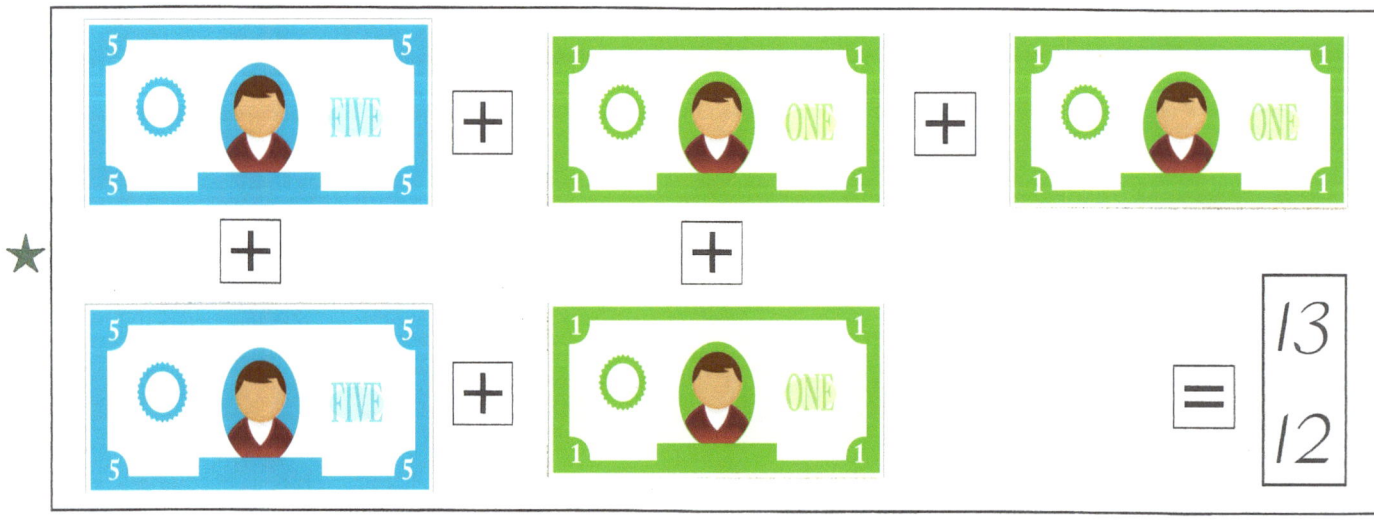

Fill in the missing numbers.

1   _   _   4   5   _   _   _   9   10

_   12   13   _   _   16   17   18   _   _

21   _   _   24   25   _   _   _   29   30

31   _   33   _   35   _   37   _   39   _

_   42   _   44   _   46   _   48   _   50

51   _   _   54   _   _   57   _   59   _

_   62   63   _   65   66   _   68   _   70

71   _   _   74   _   _   77   _   79   _

_   82   83   _   85   _   _   88   _   90

91   _   _   94   _   96   _   _   _   _

Circle the shape that comes next.

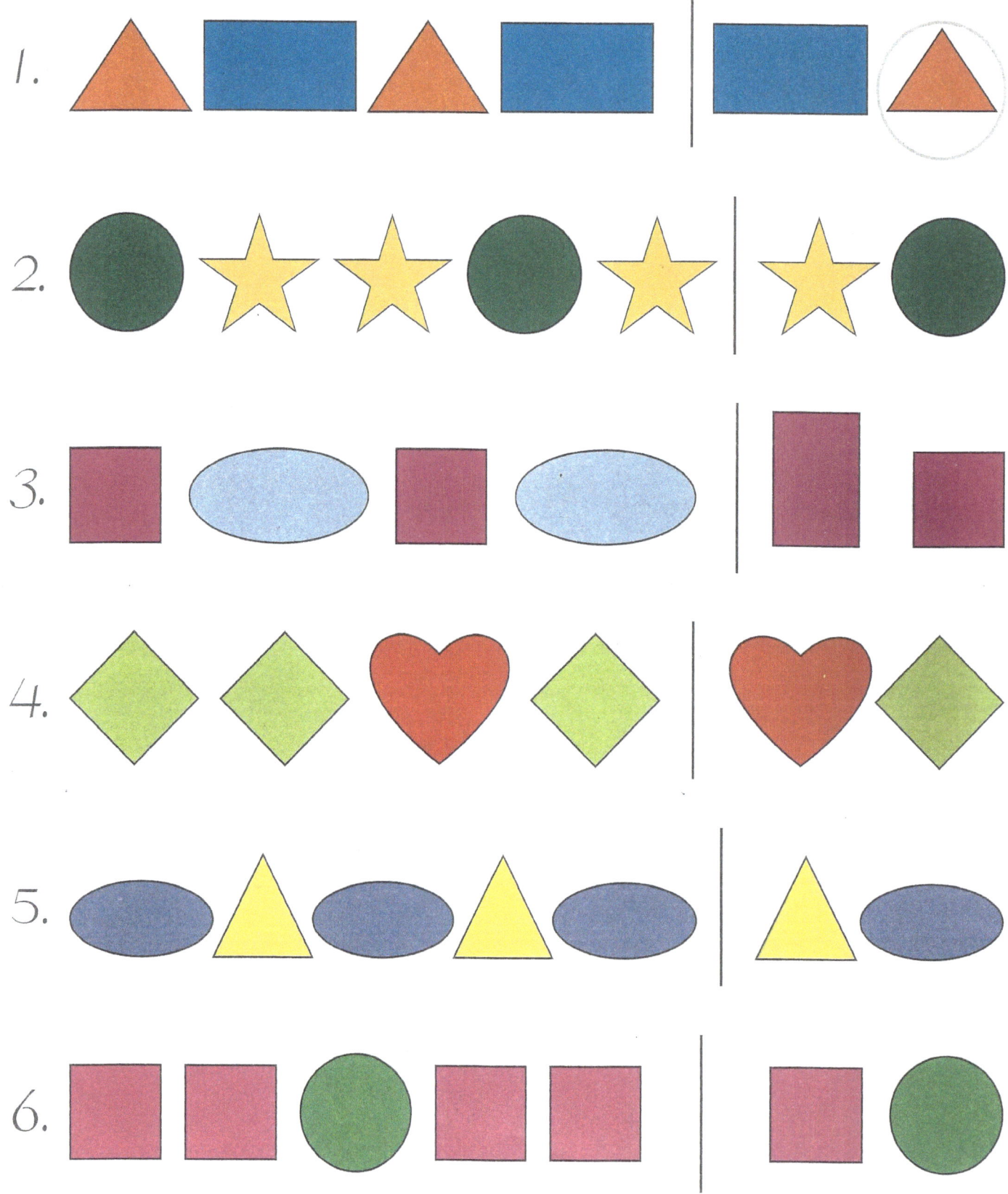

1.

2.

3.

4.

5.

6.

Add and circle the correct number.

★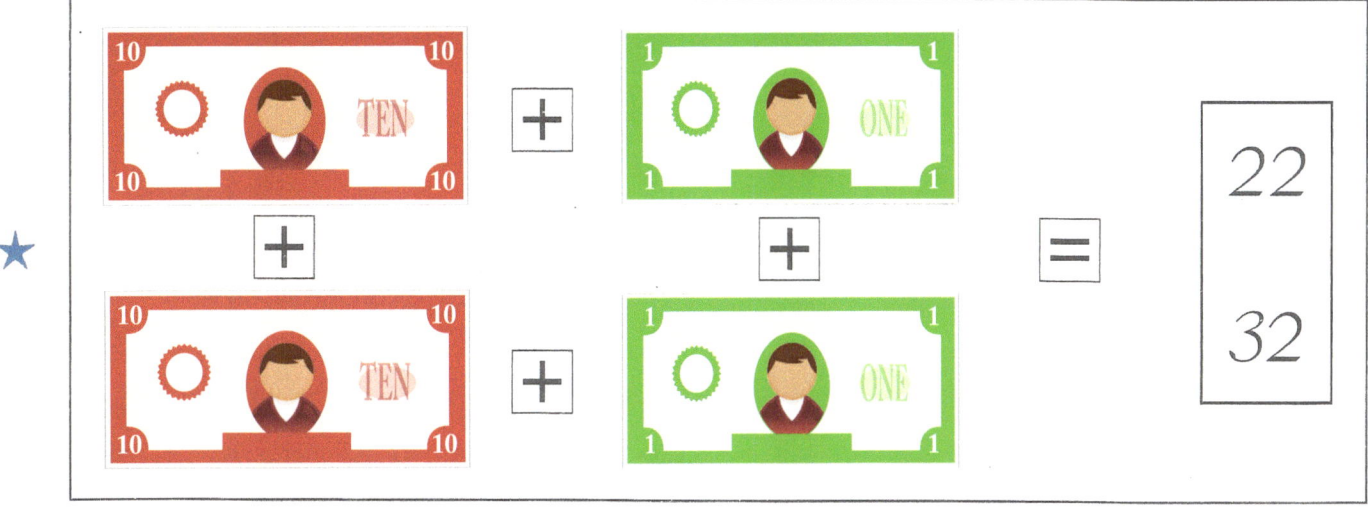

★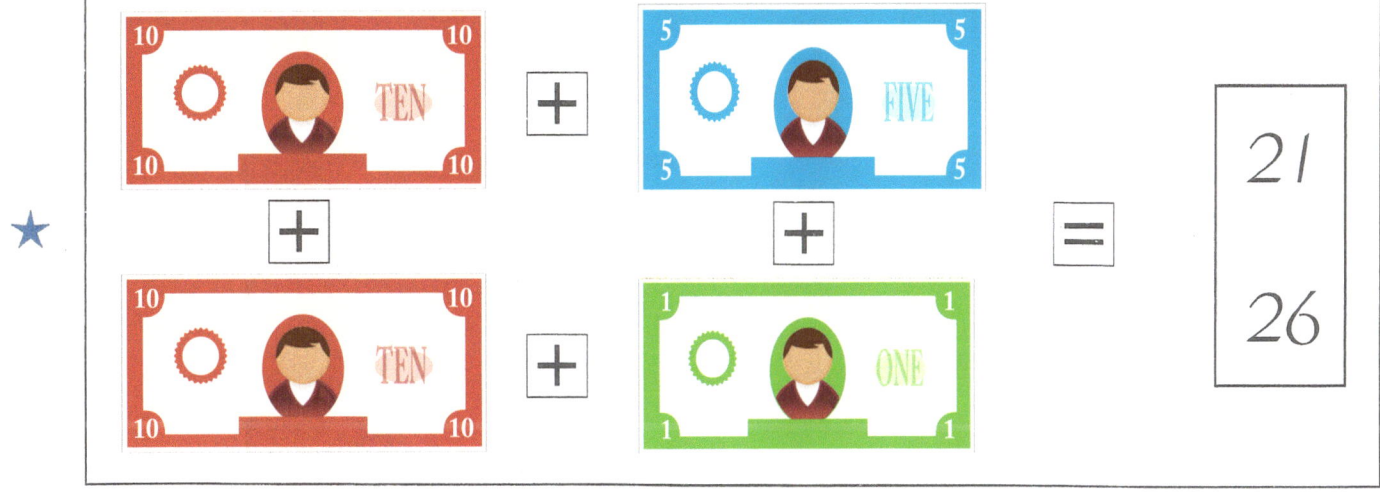

★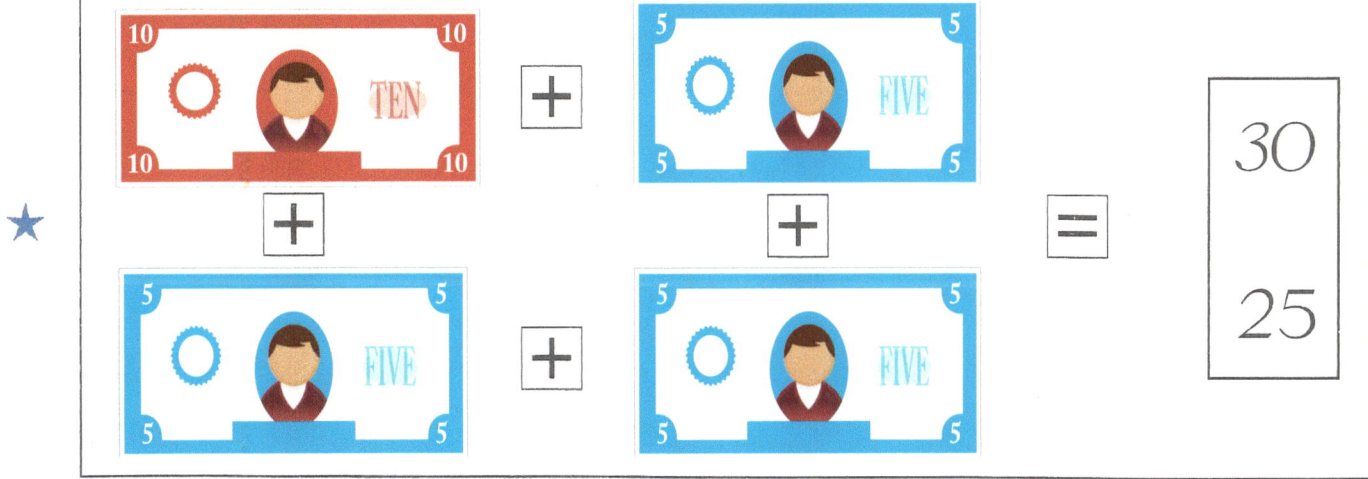

Subtract and write.

Write the number before or after.

78   ---

---   65

---   83

55   ---

---   46

90   ---

Join the dots.

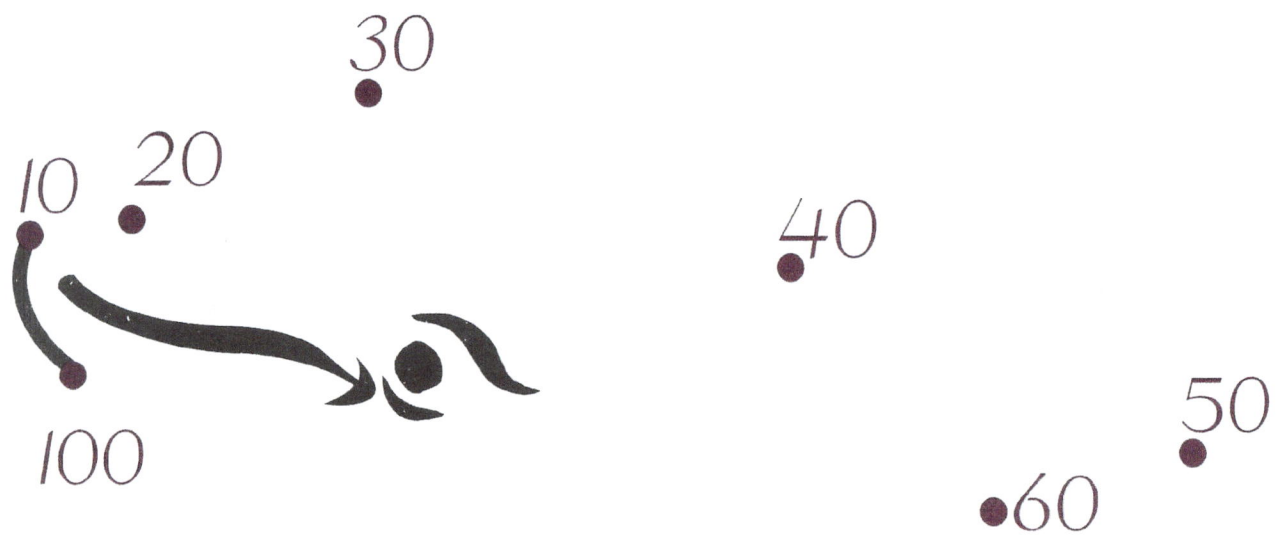

30

10   20

100

40

50

•60

•80

90

70

Revision and extension

Add and colour.

8 = blue   9 = yellow   10 = red

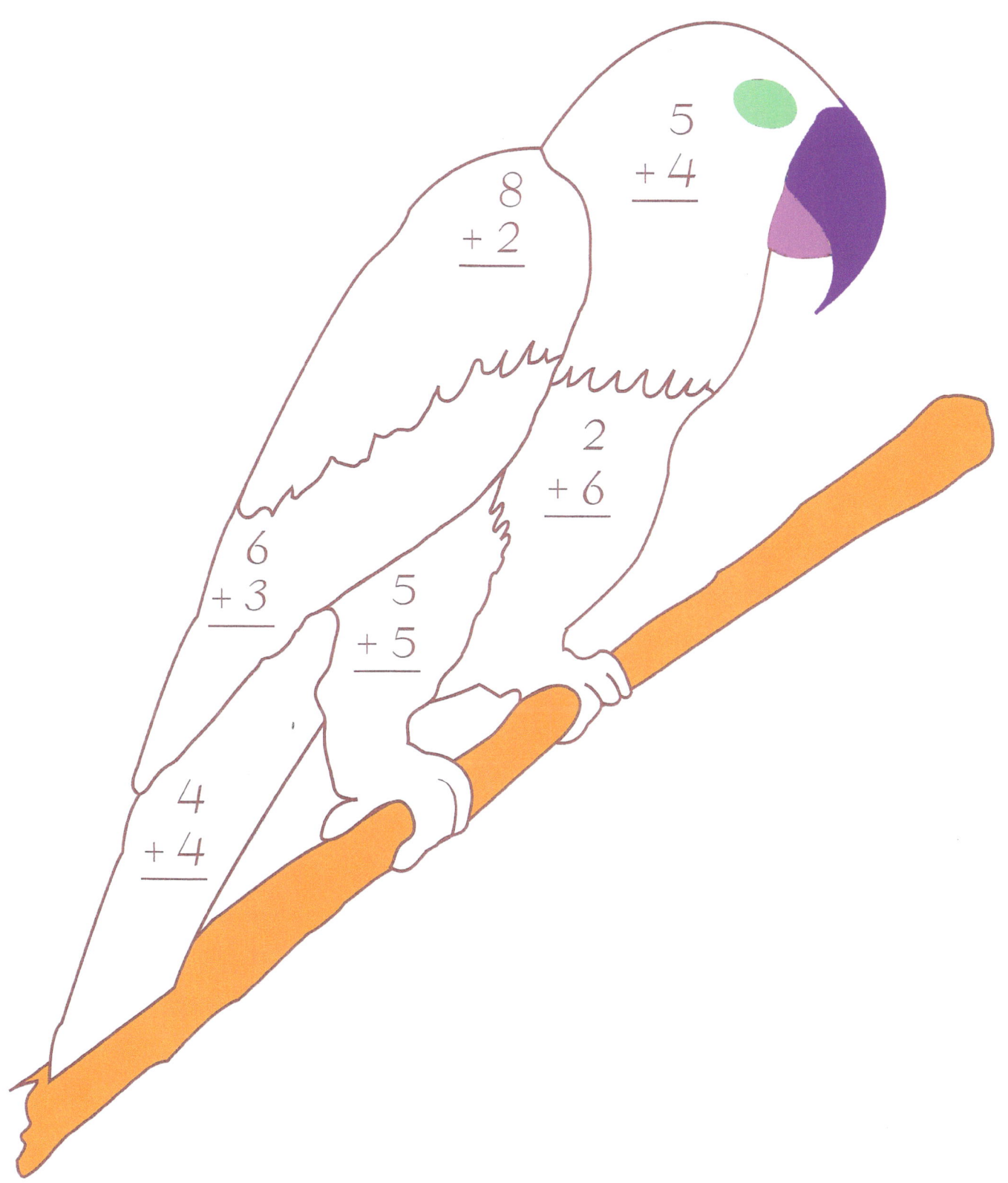

Revision and extension

Subtract, write and match.

★ 7 − 2 = 5

★ 8 − 1 = __

★ 9 − 6 = __

★ 10 − 1 = __

★ 8 − 2 = __

★ 7 − 3 = __

★ 10 − 2 = __

Revision and extension

Write the equation.

★

★

★

Revision and extension

Subtract and write.

★ $3 - 1 =$ _____   ★ $6 - 4 =$ _____

★ $5 - 2 =$ _____   ★ $10 - 1 =$ _____

★ $8 - 2 =$ _____   ★ $7 - 4 =$ _____

★ $4 - 3 =$ _____   ★ $5 - 3 =$ _____

★ $6 - 1 =$ _____   ★ $8 - 4 =$ _____

★ $10 - 2 =$ _____   ★ $10 - 3 =$ _____

★ $9 - 3 =$ _____   ★ $8 - 5 =$ _____

Evaluation

Page Number	Now I can ...	🙂	🙂	🙁	Date	Signature
Page 1	remember 1 to 10					
Page 2	play with numbers					
Page 3	remember the greater and smaller number					
Page 4	remember addition					
Page 5	name shapes recognise the same shapes					
Page 6 and 7	count to 20 match pairs					
Page 8	count to 20					
Page 9	add up to 20 write an addition equation					
Page 10	recognise objects of similar shape					
Page 11	colour according to a key					
Page 12 and 13	count to 30 match pairs					
Page 14	subtract					
Page 15	match objects and shapes					
Page 16	count to 40 add more to a group to reach a number					

Evaluation

Page Number	Now I can ...	:)	:|	:(	Date	Signature
Page 17	write a subtraction equation					
Page 18	scan pictures to find differences					
Page 19	count to 50 / add more to a group to reach a number					
Page 20	add money					
Page 21	copy a pattern					
Page 22	add money					
Page 23	count to 100					
Page 24	decide which shape comes next					
Page 25	add money					
Page 26	subtract money					
Page 27	play with numbers					
Page 28	count in tens					
Page 29	do vertical addition equations					
Page 30	subtract					
Page 31	write a subtraction equation					
Page 32	subtract					

www.ingramcontent.com/pod-product-compliance
Lightning Source LLC
Chambersburg PA
CBHW050405180526
45159CB00005B/2156